Ms. Cook

# WOOL

Jacqueline Dineen

**The story of the clothes, carpets, and furnishings whose beginnings lie on the great sheep stations of New Zealand, Australia, South Africa, and South America.**

ENSLOW PUBLISHERS, INC.
Bloy St & Ramsey Ave.
Box 777
Hillside, N.J. 07205

# Contents

*The picture above shows a wool carpet tufting machine in operation.*
*[cover] The cover picture shows shearers at work in a shearing shed in New Zealand.*
*[title page] The picture on the title page shows a flock of Blackface sheep on a farm in Scotland.*
*[1–26] All other pictures are identified by number in the text.*

*We are grateful to the International Wool Secretariat, London, for their considerable help with both text and pictures.*

This series was developed for a worldwide market.

First American Edition, 1988
© Copyright 1985 Young Library Ltd
Printed in the United States of America

10 9 8 7 6 5 4 3 2 1

Library of Congress Cataloging-in-Publication Data

Dineen, Jacqueline.
  Wool.
  (The World's harvest)
  Includes index.
  Summary: Explains how wool is removed from sheep and used in our clothes, carpets, furnishings, and many other things.
    1. Wool--Juvenile literature. [1. Wool]
I. Title. II. Series: Dineen, Jacqueline. World's harvest.
TS1547.D56 1988     677'.31     88-1187
ISBN 0-89490-227-X

# Introduction

The human race learned about wool thousands of years ago. Wandering tribes discovered how to collect wool from sheep and weave it into cloth. Today we can make cloth from other materials, but no one has found a material quite like wool. Only the sheep can provide the real thing.

There are 1,000 million sheep in the world today, and wool is a huge industry. I tell you about that industry in this book.

Today's sheep have been bred carefully to produce various sorts of wool. Some is good for making clothes, some for making carpets, some

A breed is a group of animals within a species whose mating and development are controlled. This is done so that useful characteristics (such as finer wool) will develop more strongly.

An auction is a sale held in public, where the goods are sold to the highest bidder. A buyer keeps offering a higher price until no other buyer wishes to top it.

for other purposes. In chapter 1, therefore, I describe a few of the breeds of sheep around the world, and tell you what the different types of wool are best used for.

In chapter 2 I talk about sheep-farming in the main wool-producing countries of the world. We look at the huge sheep farms in Australia, New Zealand, South Africa, and South America. The whole of picture [1] shows only a tiny corner of one of these farms. We also look at sheep in smaller countries like Britain. I tell you about herding the sheep, shearing time, and preparing the wool for sale.

The wool from the farms is sold at wool auctions. In chapter 3 we see how the buyers choose the best wool for their needs, and send it to the manufacturers in those countries which need to import it.

When the wool arrives at its destination, it goes to a woollen mill to be made into yarn and cloth. It has to be cleaned, untangled, spun into yarn, then woven or knitted into cloth. In chapter 4 I tell you about these processes, and a little about dyeing and finishing the cloth before it is made into clothes.

In chapter 5 we take a look at all the uses wool can be put to. Clothing is an important one, of course, but there are many others. I think you are going to be surprised to learn what an important part of the world's harvest wool really is.

# 1 · *What is wool?*

Today there are more than 1,000 million sheep in the world and they produce nearly 3,000 million kilogrammes of wool (6,600 million pounds) every year. Most countries have some sheep, but the most important wool-producing countries are New Zealand, Australia, South Africa, Argentina, and Uruguay. These five countries are all in the southern hemisphere, which produces 84 per cent of all the world's wool. None of these countries had sheep originally. They were all introduced from Europe and Asia, during the last few hundred years.

Sheep farming in North America, Europe, and Asia is not on the same scale, although a lot of wool is produced there too. Every country in the world needs wool. Most of them have their own sheep, and import extra wool as well.

Most animals are covered with fur or hair which is short and straight. The hair of the sheep—wool—is different from any other animal's coat. It is soft and curly, and grows quite long. It mats together easily, and so can be made into long threads. It is also fairly waterproof. Picture [2], a greatly magnified photograph of a sheep's hair, shows it to be covered with hundreds of tiny, overlapping scales which are a bit like the tiles on a roof. When water falls on the hair, it rolls off along the scales. This keeps the sheep dry during rain, as well as the clothing which is made

[2]

from wool.

People have been making clothes from sheep's wool for thousands of years. After the wool is cut off the sheep, the separate hairs are twisted together to form a continuous thread. To make cloth, these lengths of thread can be woven together in a criss-cross pattern to make a kind of web.

When people began spinning and weaving sheep's wool thousands of years ago, it was all a bit crude. Sheep were not always white or black—they could be cream, or brown, or even patchy. Nor was their wool all soft and of the same thickness—each animal could have coarse hairs mixed with the fine ones. All these different strands were woven together, and the wool was of poor quality.

In later centuries, however, people expected their clothes to look good as well as to keep them covered and warm, and they began to breed sheep to produce better types of wool. That means that the parents of a sheep are not allowed to choose each other—the best sheep are selected by the farmer to produce offspring which inherit the characteristics he requires. Within a few generations a farmer can produce a whole flock of the type he wants.

[3]

## The importance of breeding

Farmers bred sheep to produce either a coarse or a fine coat, not a mixture. They bred them to produce wool of a pure and even colour instead of patches of different colours. Today the farmers and the buyers know what kind of wool they are going to get simply by knowing

what breed of sheep it has come from. This is
important because various types of wool have
different uses. Soft wool, for example, is used
for clothing; coarse wool is used for carpets.

The most important sheep breed in the
world is the Australian Merino, which you see
in picture [3]. It has a thick, heavy fleece and
the wool is of very good quality. The Merino
came from Spain in the nineteenth century,
but now it is Australia's main breed. Almost
three-quarters of the sheep in Australia are
Merinos. Because their wool is both soft and
strong, it is used for fine clothing and also for

[ 4 ]

7

[5]

things which need to be hard-wearing.

Many of the finest sheep breeds came from Britain. One group of them are known as British Longwools. Longwools are now found all over the world and their wool is used for carpets, clothing, and knitting yarns. The Longwools are named after the area of Britain from which they first came. For example the Romney, which you see in picture [4], came from Romney Marsh in Kent. It is now New Zealand's main breed. The Longwool's fleece is coarser than the Merino's, and the sheep can survive comfortably on poor land.

The Shortwools, also from Britain, are found in all the main sheep-producing countries. The sheep are fairly small, and give fine, short wool. Picture [5] shows a Shortwool called the Southdown (from the South Downs in

[6]

England) which is another important breed in New Zealand.

In South Africa the Karakul is a popular breed. The Karakul came from Asia, and its coarse wool is very suitable for carpets. You can see a pair of Karakul sheep in picture [6]. Their wool is normally left in its natural colours during manufacture, and not dyed as most wool is.

Farmers in the main wool-producing countries have, over the years, produced new breeds of sheep. The Corriedale, for example, was produced by cross-breeding the Merino with British Longwool breeds. This sheep is now found in large numbers in New Zealand, Australia, and South America. It is heavier than the Merino, and its wool is a bit coarser.

I should just mention, before we reach the end of this chapter, that sheep are not kept only for wool. Sheep provide a very important source of meat in most countries of the world. However, the sheep raised for meat are usually different breeds from those kept for wool.

# 2 · Sheep farming

To graze means to feed on growing grass. It is the main, or only, source of food for sheep, cows, goats, horses, and other animals.

A herd is a number of sheep, or similar animals, who are kept together. Herding means gathering them or keeping them travelling together, perhaps to new pastures or to the annual shearing.

In many countries sheep are only one part of a farm. There may be cows, pigs, poultry, and even crops as well. The sheep are left to graze in the fields, or on heath or moorland, for most of the year. Their lambs are born in the spring. The sheep are shorn once a year, in the spring or early summer. On the small farms of Europe the sheep can be herded by one farmer and his dog.

However, in the main wool-producing countries, there are huge farms which have nothing but sheep. In Australia and New Zealand these farms are known as sheep stations. In Australia a typical station has 3000 or 5000 sheep and no other animals or crops at all. The very large sheep stations might have as many as 100,000 in a flock. The stations are so enormous that the sheep may be many miles away. They have to be herded by a group of men on horses or motorbikes. Just imagine how the man in picture [7] would manage without his horse. There are about 80,000 sheep stations in Australia, and 135 million sheep.

New Zealand also has huge sheep farms. There are about 30,000 sheep stations in New Zealand, and 68 million sheep. There are twenty times more sheep than people! New Zealand wool is fairly coarse and very strong, and is the best carpet wool in the world.

South Africa has a very varied climate. Parts of the country are ideal for crops, but there are large areas which are almost desert, and

[8]

picture [8] shows sheep being watered in a drought.

A drought is a long period without rain, causing grass and other plants to wither, rivers to dry up, and animals to suffer from lack of drinking water.

There are about 30 million Merinos in South Africa and they will thrive on poor grass and dry soil; but however tough they are, they cannot survive for long without water.

Argentina and Uruguay are the two main wool-producing countries of South America. Those countries, also, have vast sheep stations. There are about 35 million sheep in Argentina, and 18 million in Uruguay.

A gaucho is (strictly) a native cowboy of mixed Spanish and South American Indian blood, but the term often covers any cowboy or herdsman of the South American plains.

On South American sheep farms the enormous flocks are herded by men on horseback called gauchos. In Uruguay and in northern Argentina the land is good, and the farms also have cattle, and grow crops such as wheat and maize. But further south, where there is hardly any rain, the grass is sparse.

The sheep have to wander over great areas to find enough to eat, and the gauchos have a long journey whenever it is necessary to round them up.

Other countries also produce large quantities of wool. The Soviet Union produces 462 million kilogrammes a year, and the United States produces 47 million kilogrammes. However, because they have such huge populations, they still have to import wool. Even Britain, which produces 51 million kilogrammes a year for a much smaller population, has to import wool.

You can see that wool is produced mainly in the large, empty countries of the southern hemisphere, and exported to the overpopulated [9]

countries of the northern hemisphere.

Whatever the size of the farm, the sheep's way of life is much the same. All sheep roam around looking for fresh grass to graze upon. If they are in an enclosed field their grazing area is small but probably has good rich grass. However, sheep will also graze happily on rough land like moor and heath where the grass is poor, and many farmers leave them to roam wherever they like. On moorland in countries like Britain they can often be seen grazing at the roadsides.

## Shearing the sheep

Once a year, between spring and autumn, sheep are rounded up and brought close in to the farm for shearing. This is usually done in large shearing sheds like the one in picture [9], or in the picture on the cover. It needs a lot of skill to shear a sheep properly because the shearer has to remove the whole fleece in one piece. An expert shearer like the New Zealander in picture [10] needs only about five minutes to shear a sheep, and can shear 125 in a day.

The shearer takes the sheep by the forelegs, and drags it backwards towards the bay where his electric shears hang from the ceiling. Holding the sheep firmly against his legs with one arm, he uses long, smooth strokes of the shears to take off the wool. He shears close to the body, but not so close as to cut the skin. The sheep will often try to wriggle, and the man has to turn the sheep about several times during the operation, holding it firmly all the time, while he works continually with the other

[10]

hand. It needs a lot of strength, especially as he is bent double all the time.

When the fleece has been removed it is laid out on a table. It is examined carefully, and any shabby bits of fur are cut off. The farmer sorts the fleeces into piles of the same type and quality. Then the fleeces are sewn into large sacks called bales. What happens to them I will tell you in the next chapter.

The shorn sheep in picture [11] look almost naked. However, they are probably very glad to be rid of their heavy fleeces, which will grow again well before the cold weather returns. The sheep will be ready for shearing again by the following spring.

In New Zealand and some other countries the sheep are shorn twice a year. The wool from the second clipping is short, and is used mainly for carpets.

[11]

A fleece is the coat of wool on a sheep, or the whole coat removed in one piece by shearing.

# 3 · *The wool industry*

When the wool leaves the farm, it is sold to manufacturers who will make it into all sorts of woollen products. In Australia, New Zealand, and South Africa the selling is done at public auctions. The auction is held at a place called a wool exchange. The exchange is usually near a port so that the wool can be shipped abroad as soon as it has been bought.

Manufacturers from all over the world send their buyers to the auctions. The buyers' job is to choose the wool, and buy it for the best possible price.

Naturally the buyers wish to inspect the wool before the auction begins. One way to inspect the wool is for the buyer to take samples from opened bales in the wool broker's store. All the bales are laid out in rows, and the buyers walk about comparing quality and texture.

This method is still used in many countries, but Australia, New Zealand, and South Africa have introduced a new system to make the inspection easier. You can see this system in action in picture [12]. Before the auction begins, samples are taken from the middle of each bale by a mechanical grab. These 'grab samples' are examined in a laboratory, which produces a certificate recording the thickness of the fibres and the weight of the wool after cleaning. In this way the buyer has some facts to rely on without having to estimate them for himself.

A broker is a person who is employed to buy or sell goods for someone else, without ever owning them. In this case, it is the sheep farmer who owns the wool, and employs the broker to sell it at the auction.

However, judging wool is still a highly
skilled job, for the buyers have to choose
between so many types. A carpet manufacturer
needs coarse wool, while a knitwear firm is
looking for fine, soft wool; but there is more to
it than that. The buyer has to look at the
colour, and judge the condition of the wool.
The length of the fibres is also important. He
has to see if the wool is reasonably clean and
free from grass, dust, burrs, and thistles. Wool
is sold by weight; so unless the exchange is
operating the 'grab sample' system the buyer

[12]

[13]

must estimate what the wool will weigh after it has been washed. When he has found the type of wool he wants, he decides what it is worth and how much he will bid for it.

In Britain, sheep farmers must send all their wool to the British Wool Marketing Board. No one else is allowed to sell British wool. The Board sorts the fleeces into grades according to their quality. It also states the type of end product they are most suitable for—knitting wools, carpets, tweeds, and so on. The grading tells the buyers exactly what they are getting, and the farmers are paid for the wool according to the grade.

## The wool-consuming countries

When the buyer has chosen all the wool he needs, and bought it, he arranges for it to be delivered to the manufacturers. In picture [13] you can see wool sewn into bales, and in picture [14] the bales are being loaded into a container. This wool, from Australia, will be put on a ship and taken to South Korea.

The countries which import wool are called wool-consuming countries. The main wool-consuming areas are Japan, eastern Europe, western Europe, and North America.

Japan imports more wool than any other country. There are only 11,000 sheep in Japan but there are more than 100 million people to clothe. Japan imports more than 170 million kilogrammes of wool each year.

Eastern Europe (including Russia, Poland, Czechoslovakia, and Yugoslavia) is the largest wool-using area in the world, but only the second-largest importer. This is because Russia produces a lot of its own wool.

The United States imports half of the wool and wool products which it needs. Canada produces only 1 million kilogrammes each year,

[14]

[15]

so has to import most of its needs. So, as I said on page 13, you can see that wool is produced chiefly in the southern hemisphere, and consumed chiefly in the northern hemisphere.

So, as I said on page 13,

*Importing wool, exporting clothes*

Many of the wool products bought by North America and Europe are not bought directly from the countries which produce wool. The wool is bought by Asian countries, who make clothing from it, then sell the clothing to the wealthier countries of the West. Picture [15] shows women hard at work in a knitwear factory in Taiwan. In Hong Kong and South Korea, also, there is a large knitwear industry.

However, the local people do not wear much wool in their hot climate. The knitwear is made for export, and sold to countries with cooler climates. Next time you buy a jumper, take a look at the label to see where it was made.

The wool trade in Britain, France, and Italy has been well established for centuries. These countries produce cloth and knitwear of fine quality. However, their own sheep produce rather coarse wool, so all three countries have to import fine wool to make these goods.

India has 40 million sheep, and also imports a large amount of wool for carpet-making and for knitwear. Many of India's fine carpets are made for export to North America and Europe, and most of its knitwear is sold to Russia and to eastern Europe.

We do not know how many sheep there are in China, but there are probably 100 million or more. Even so, China is a major importer of wool from Australia and New Zealand.

# 4 · *At the woollen mill*

A mill is a building containing machinery for the spinning or weaving of cotton or wool (or for the grinding of corn into flour, and similar operations).

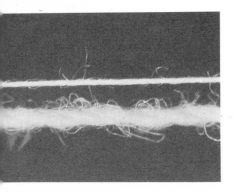

[16]

The first job of the mill is to turn the raw wool from the sheep's back into yarn. Yarn is the long woollen thread which you see in shops, rolled into balls or wound on to spools. It is the basic material used for nearly all woollen products whether they are knitted or woven.

There are two methods of making yarn. One is called the woollen process, and the other is called the worsted process. The main difference between the two is that worsted yarn is smoother than woollen yarn. This is because the wool for worsted yarn is combed so that the short fibres are removed and the long fibres all lie in the same direction. In woollen yarn, the short fibres are left, and the fibres do not all lie in the same direction, so that woollen yarn is bulky and has a fuzzy surface. Picture [16] shows the difference between worsted (at the top) and woollen yarn.

Wool with long fibres is best for making worsted yarn, and wool with short fibres is best for making woollen yarn.

Worsted cloth is made into garments like men's suits, which have to look smooth and crisp. Woollen yarns are used for knitting, and also for weaving into fairly coarse fabrics like tweed.

The woollen process is carried out all at one mill, from raw wool to finished fabric, but wool for worsted may go to several mills during its manufacture. There are separate mills for combing, spinning, weaving, dyeing, and finishing.

When the wool arrives at the mill it contains grease and dirt from the sheep's body, and grass and dust from the fields. The wool is sold in this state, so this is how it arrives at the mill. The process of washing is called scouring. The wool is passed through a series of washtubs containing soapy solutions. The natural grease (called lanolin) is removed from the wool, and so are all the bits of grass and dirt. Picture [17]

*Cleaning and spinning*

[17]

[18] shows the wool being scoured. Washed wool can weigh 30 per cent less than greasy wool. This is why the buyers have to use all their skill and knowledge at the auctions.

You may have heard of lanolin before. The lanolin from the wool is not wasted, but used to make cosmetics and soap.

If the wool contains too many burrs and seeds to remove by scouring, it is passed through machinery which dries and crushes the burrs so that they can be shaken out.

Scouring causes the fibres to tangle, so the next stage is carding. You can see a row of carding machines in picture [18]. The wool is passed over revolving cylinders and rollers with wire teeth. This untangles the fibres, and produces thin ropes of wool called slivers.

If the wool is to be used for worsted yarns,

the slivers are now combed to remove the short fibres, and make the long fibres all lie the same way to form smooth ropes. Picture [19] shows these ropes wound into balls called tops, which can be spun into worsted yarn.

So both lots of wool are now ready for spinning into yarn. It does not matter whether they are the more tangled fibres used for the woollen process, or the combed fibres used for worsted—the spinning process is the same for both.

The spinning process has come a long way since the early method of twisting it into yarn by hand. It is the twisting which joins all the separate strands of wool into one long, continuous length of yarn. In the modern mills the wool is twisted on large and complicated machines called spinning machines. If you look at a piece of wool, you can see how it consists of thin strands all twisted together. If you untwist the wool, you are left with the separate strands. In fact you can untwist a length of wool right back to the small fibres it is made of.

Some wools are used in their natural colour, but most are dyed. Woollen yarns are usually dyed before spinning. Worsted yarns may be dyed before spinning, but they are often dyed after the wool has been spun into yarn. Picture [20] shows wool dyed before it has been spun, and picture [21] shows wool dyed after it has been spun.

Now that the yarn has been produced, it must be made up into cloth. The two basic methods of producing cloth are weaving and knitting.

[19]

[20]

*Weaving and knitting*

[21]

Weaving is done on a machine called a loom like the one in picture [22]. The cloth is made by criss-crossing lengths of yarn into a sort of web. One set of yarns, which passes from end to end of the web, is called the warp. The yarn which crosses and intersects the warp threads at right angles is called the weft. The weaver can make patterns in the cloth by varying the colour of the threads, or by varying the pattern of the weave. The basic weaving pattern is for the weft thread to go over and under each warp thread in turn, but it could go over two and under one, or over three and under five, or in any other pattern. Next time you see a cloth which looks as if it is made of only a single colour, look closer! You will probably find it contains several colours and patterns.

[22]

You have all seen women knitting on knitting needles. Perhaps you can knit, too. However, there are also large machines which can knit, and you can see one of them in picture [23]. The principle is just the same. The cloth is made from a single, continuous length of yarn by forming it into a series of connecting loops. It is really very much easier to watch someone doing it than to have it explained in words!

Knitted fabrics can be made in a plain stitch in one colour, or in a variety of patterns and colours.

Whether woven or knitted, the fabric is now ready to be made into clothing. First, however, it must be 'finished'.

It is important that the cloth keep its shape when it is made up into clothes. It must be smooth and soft and comfortable to wear. The finishing process adds these final touches to the cloth. Woollen fabric is brushed to give it a soft texture, or pressed to give it a glossy surface.

There are two basic types of knitting machine—flatbed, for making lengths of fabric, and circular, for making a tube of fabric for garments like jerseys and socks. The machine in picture [23], however, is neither of these. It is a rib machine, for making collars, cuffs, and welts.

# 5 · *How wool is used*

The woollen and worsted fabrics are now ready to leave the woollen mills. What will the cloth be used for?

You can probably think of plenty of knitted clothes—sweaters, scarves, gloves, socks, hats. This is one use for wool. You probably have clothes made of woollen or worsted cloth—coats, skirts, suits, trousers. That is another use. The clothing industry is by far the biggest consumer of wool.

## The carpet industry

Another important use which I have already mentioned is carpet-making. The best carpets are made from wool, and they are made on looms just as cloth is. Some of the most famous and expensive carpets in the world come from India and Iran, made on old wooden looms worked by hand. Carpets like these can last for hundreds of years. They are usually woven in distinctive designs of glowing colours and named after the area they come from.

However, not everyone can afford individually-made carpets. It is more usual to make carpets on machine looms, which can produce several hundred of the same pattern. The towns of Axminster and Wilton in England have given their names to the type of weave used to make carpets there. Picture [24] shows part of an Axminster loom in operation.

Even machine-made woven carpets are expensive. A much cheaper process is the

[24]

'tufted' carpet. This consists of very short pieces
of wool, perhaps a centimetre in length,
threaded around a prepared net.

Wool is used to make many other household
items, such as blankets, chair covers, curtains,
and fillings for quilts, duvets, and sleeping
bags. Look around your home and see how
many things have been made with wool.

Wool is also used in more unusual ways.
Take a look at picture [25]. Wool can be
treated to make protective clothing for people
like firemen and racing drivers. This is because

[25]  wool does not catch fire, as some man-made
fibres do, it merely smoulders and chars.

So, with all the advantages of wool, why
bother with man-made fibres? Well, many of
these fibres were developed because there was
not enough wool to go round. Man-made
fibres were produced as substitutes for wool.
However, it was soon discovered that they had
some advantages over wool. They are usually
cheaper, they are easier to wash, and some of
them are less likely to shrink or stretch.

As people began to turn to other fabrics in
the 1920s and 1930s, sheep farmers thought
that the true quality of wool might be
forgotten. So the International Wool
Secretariat was set up in 1937. The IWS is
funded by five of the main wool-producing
countries—New Zealand, Australia, South

Africa, Brazil, and Uruguay. It promotes the use of wool, and carries out research into ways of making it more suitable for modern needs.

The IWS introduced the 'Woolmark' in 1964, to avoid confusion between wool and man-made fibres. A garment or fabric with this symbol on the label is made of 'pure new wool'. 'Pure' means that the wool is not mixed with other fibres. 'New' means that the wool is being used for the first time. The blankets in picture [26] carry the Woolmark.

Sometimes it is an advantage to add a small proportion of another fibre to wool, for example to give it a lighter weight. The 'Woolblendmark', introduced in 1971, shows that a garment contains at least 60 per cent new wool. The two symbols are now used all over the world.

[26]

# Index

**Acknowledgements for photographs:** *All photographs are kindly provided by The International Wool Secretariat, London, except the one on the cover and nos. 1 and 10 which are provided by National Publicity Studios, Wellington, New Zealand.*